I0468022

NUREG-0090
Vol. 28

Report to Congress on Abnormal Occurrences

Fiscal Year 2005

Date Published: April 2006

Division of Fuel, Engineering and Radiological Research
Office of Nuclear Regulatory Research
U.S. Nuclear Regulatory Commission
Washington, DC 20555-0001

ABSTRACT

Section 208 of the Energy Reorganization Act of 1974 (Public Law 93-438) defines an "abnormal occurrence" (AO) as an unscheduled incident or event that the U.S. Nuclear Regulatory Commission (NRC) determines to be significant from the standpoint of public health or safety. The Federal Reports Elimination and Sunset Act of 1995 (Public Law 104-66) requires that the NRC must report AOs to Congress annually. This report describes those events that the NRC or an Agreement State identified as AOs during fiscal year (FY) 2005.

The report describes three events at NRC-licensed facilities that meet the criteria to be classified as AOs, as defined in Appendix A to the report. All three events occurred at medical institutions. The first event involved a patient who received the incorrect dose distribution while undergoing therapeutic brachytherapy[1] treatment. The second event involved an infant who was administered the incorrect diagnostic dosage of technetium-99m. The third event involved three patients who received unintended radiation doses to the skin of their thighs while undergoing therapeutic treatment.

Reports from Agreement States are also included. Agreement States are those States that have entered into formal agreements with the NRC pursuant to Section 274 of the Atomic Energy Act (AEA) to regulate certain quantities of AEA material at facilities located within their borders. Currently, there are 34 Agreement States (Minnesota became the 34th Agreement State on March 31, 2006). During Fiscal Year 2005, Agreement States reported six events that occurred at Agreement State-licensed facilities, including five therapeutic medical events and one diagnostic medical event. All six events met the criteria for AO categorization.

Appendix A to this report presents the NRC's criteria for selecting AOs, as well as the guidelines for selecting "Other Events of Interest." Appendix B, "Updates of Previously Reported Abnormal Occurrences," does not contain any updated information on AO events reported in the FY 2004 Report to Congress on Abnormal Occurrences because no new significant information became available. Appendix C, "Other Events of Interest," contains one new event of interest on safe-shutdown safety-related systems at the Kewaunee Power Station and updated information on a spent fuel record accountability discrepancy at the Humboldt Bay Power Plant.

---

[1]Brachytherapy means a method of radiation therapy in which sources are used to deliver a radiation dose at a distance of up to a few centimeters by placement of sources on the body surface, in natural body cavities, or by placement directly in tissues.

CONTENTS

PREFACE

INTRODUCTION

Section 208 of the Energy Reorganization Act of 1974 (Public Law 93-438) defines an "abnormal occurrence" (AO) as an unscheduled incident or event that the U.S. Nuclear Regulatory Commission (NRC) determines to be significant from the standpoint of public health or safety. The Federal Reports Elimination and Sunset Act of 1995 (Public Law 104-66) requires that the NRC must report AOs to Congress annually. This report describes those events that the NRC or an Agreement State identified as AOs during fiscal year (FY) 2005. [Agreement States are those States that have entered into formal agreements with the NRC pursuant to Section 274 of the Atomic Energy Act (AEA) to regulate certain quantities of AEA material at facilities located within their borders].

For the purpose of this report, the NRC defined AOs using the criteria set forth in Appendix A. The NRC initially promulgated those criteria in a policy statement that the Commission published in the *Federal Register* on February 24, 1977 (42 FR 10950), followed by several revisions in subsequent years.

The NRC has determined that, of the incidents and events reviewed for this reporting period, only those that are described herein meet the criteria for being reported as AOs. The information reported for each AO includes the date and place, nature and probable consequences, cause(s), and actions taken to prevent recurrence.

Appendix A to this report presents the NRC's criteria for selecting AOs, as well as the guidelines for selecting "Other Events of Interest." Appendix B, "Updates of Previously Reported Abnormal Occurrences," does not contain any updated information on events reported in the FY 2004 Report to Congress on Abnormal Occurrences. Appendix C, "Other Events of Interest," presents information concerning events that are not reportable to Congress as AOs, but are included in this report based on the Commission's guidelines, as listed in Appendix A to this report. Specifically, this appendix contains one new event of interest on safe-shutdown safety-related systems at the Kewaunee Power Station and updated information on a record accountability discrepancy at the Humboldt Bay Power Plant.

To widely disseminate information to the public, the NRC issues *Federal Register* notices describing AOs at facilities licensed or otherwise regulated by the NRC or an Agreement State. Information on activities licensed by Agreement States is also publicly available from the Agreement States.

THE REGULATORY SYSTEM

The system of licensing and regulation by which the NRC carries out its responsibilities is implemented through the rules and regulations in Title 10 of the *Code of Federal Regulations* (10 CFR). Public participation is one essential element of the regulatory process. To accomplish its objectives, the NRC regularly conducts licensing proceedings, inspection and enforcement activities, operating experience evaluations, and confirmatory research. In addition, the NRC maintains programs to establish standards and issue technical reviews and studies.

The NRC adheres to the philosophy that the health and safety of the public are best ensured by establishing multiple levels of protection. These levels are normally achieved and maintained through regulations specifying requirements that ensure the safe use of radioactive materials. Those regulations contain design and quality assurance criteria appropriate for the various activities regulated by the NRC. Licensing, inspection, and enforcement programs provide a regulatory framework to ensure compliance with regulations. The NRC is striving to make the regulatory system more risk-informed and performance-based, where appropriate.

REPORTABLE OCCURRENCES

Review and response to operating experience are essential to ensure that licensed activities are conducted safely. Toward that end, the regulations require that licensees must report certain incidents or events to the NRC. Such reporting helps to identify deficiencies and ensure that corrective actions are taken to prevent recurrence.

The NRC and industry review and evaluate operating experience to identify safety concerns. Information from the review and evaluation is disseminated and fed back to licensees through licensing activities and regulations. Operational data are maintained in computer-based data files for more effective collection, storage, retrieval, and evaluation.

Except for records that statutes or regulations exempt from public disclosure, the NRC routinely disseminates information on reportable occurrences at licensed or regulated facilities to the industry, the public, and other interested groups when the occurrences happen. This dissemination is achieved through public announcements and special notifications to licensees and other affected or interested groups. In addition, the NRC routinely informs Congress of significant events occurring at licensed or regulated facilities.

AGREEMENT STATES

Section 274 of the Atomic Energy Act, as amended, authorizes the Commission to enter into agreements with States whereby the Commission relinquishes, and the States assume, regulatory authority over byproduct, source, and special nuclear materials in quantities not capable of sustaining a chain reaction. States who enter into such agreements with the Commission are known as Agreement States. Agreement States must maintain programs that are adequate to protect public health and safety and are compatible with the Commission's program for such materials. During Fiscal Year 2005, there were 33 Agreement States. On March 31, 2006, Minnesota became the 34th Agreement State.

In early 1977, the Commission determined that events that meet the criteria for AOs at facilities licensed by Agreement States should be included in the quarterly report to Congress. Therefore, AOs reported by the Agreement States to the NRC are included in the AO report and in the *Federal Register* notice issued to disseminate the information about each AO to the public. Agreement States report event information to NRC in accordance with compatibility criteria established by the "Policy Statement on Adequate and Compatibility of Agreement State Programs," published in the *Federal Register* notice on September 2, 1997 (62 FR 46517). Procedures have been developed and implemented for evaluating materials events to determine those that should be reported as AOs. The AO criteria in Appendix A are applied uniformly to events at facilities regulated by the NRC and the Agreement States.

FOREIGN INFORMATION

The NRC exchanges information with various foreign governments that regulate nuclear facilities. This foreign information is reviewed and considered in the NRC's research and regulatory activities, as well as its assessment of operating experience. Although foreign information may occasionally be referred to in the AO reports to Congress, only domestic AOs are reported.

UPDATES OF PREVIOUSLY REPORTED ABNORMAL OCCURRENCES

The NRC provides updates of previously reported AOs if significant new information becomes available. These updates appear in Appendix B to this report.

OTHER EVENTS OF INTEREST

The NRC provides information concerning events that are not reportable to Congress as AOs, but are included in this report based on the Commission's guidelines, as listed in Appendix A. Other events of interest appear in Appendix C to this report.

ACRONYMS and ABBREVIATIONS

AEA	Atomic Energy Act
AO	abnormal occurrence
Bq	becquerel
CFR	*Code of Federal Regulations*
cGy	centigray
Ci	curie
cm	centimeter
Cs-137	cesium-137
FR	*Federal Register*
FY	Fiscal Year
GBq	gigabecquerel
Gy	gray
HDR	high dose-rate
I-123	iodine-123
I-125	iodine-125
I-131	iodine-131
Ir-192	iridium-192
in	inch
MBq	megabecquerel
μCi	microcurie
mCi	millicurie
NRC	U.S. Nuclear Regulatory Commission
SNM	special nuclear material
Sr-90	strontium-90
Sv	sievert
Tc-99m	technetium-99 metastable
TEDE	total effective dose equivalent
Y-90	yttrium-90

ABNORMAL OCCURRENCES
IN FISCAL YEAR 2005

NUCLEAR POWER PLANTS

During this period, no events at U.S. nuclear power plants were significant enough to be reported as AOs.

****

FUEL CYCLE FACILITIES
(Other Than Nuclear Power Plants)

During this period, no events at U.S. fuel cycle facilities were significant enough to be reported as AOs.

****

OTHER NRC LICENSEES
(Industrial Radiographers, Medical Institutions, etc.)

During this reporting period, three events at NRC-licensed or regulated facilities were significant enough to be reported as AOs based on the criteria in Appendix A to this report.

05-01 Medical Event at the University of Minnesota in Minneapolis, Minnesota

Criterion IV, "For Medical Licensees," of Appendix A to this report states, in part, that a medical event that results in a dose that is (1) equal to or greater than 1 Gy (100 rads) to a major portion of the bone marrow, to the lens of the eye, or to the gonads or (2) equal to or greater than 10 Gy (1,000 rads) to any other organ; and represents a prescribed dose or dosage that is delivered to the wrong treatment site will be considered for reporting as an AO.

Date and Place — January 24, 2005, Minneapolis, Minnesota

Nature and Probable Consequences — The licensee reported that a patient being treated for cervical cancer received an incorrect dose distribution. One area of the cervix received 8.21 Gy (821 rads) instead of the intended 16.43 Gy (1,643 rads). Another area of the cervix received 3.72 Gy (372 rads) instead of the intended 4.65 Gy (465 rads). Additionally, other locations received higher than intended doses. The intended doses to the bladder and the rectum were 11.47 Gy (1,147 rads) each, but they received 14.48 Gy (1,448 rads) and 20.12 Gy (2,012 rads), respectively. The treatment involved an applicator with an insert which contained low-dose radiotherapy sources. The licensee cut the insert 6 centimeters (cm) too short so that when the applicator was positioned in the patient's cervix, the three cesium-137 (Cs-137) sources were not extended the proper distance. The referring physician and patient were informed of this event. The licensee does not believe that this event will have any adverse health effects on the patient. The patient subsequently received a follow-up treatment to deliver the full intended dose to the treatment sites.

Cause(s) —This event was caused by human error. The incorrect dose was administered to the incorrect location.

Actions Taken to Prevent Recurrence — Corrective actions taken by the licensee included stopping all low dose-rate treatments until all individuals are trained, and modifying their procedures to incorporate a dual verification system.

This event is closed for the purpose of this report.

****

05-02 Medical Event at St. Johns Mercy Hospital in St. Louis, Missouri

Criterion I.A.2, "For All Licensees," of Appendix A to this report states, "Any unintended radiation exposure to any minor (an individual less than 18 years of age) resulting in an annual total effective dose equivalent (TEDE) of 50 millisieverts (mSv) (5 rem) or more, or to an embryo/fetus resulting in a dose equivalent of 50 mSv (5 rem) or more," will be considered for reporting as an AO.

Date and Place — March 9, 2005, St. Louis, Missouri

Nature and Probable Consequences — The licensee reported that a 5-month old infant was prescribed 18.5 MBq (0.5 mCi) of technetium-99 metastable (Tc-99m), but instead received 414.4 MBq (11.2 mCi) of Tc-99m. Hospital personnel did not look at the dosage label to verify the dose to be administered. The whole body dose to the infant was calculated to be between 0.052 to 0.10 Sv (5.2 to 10 rem). The physician informed the infant's parents. The NRC's medical consultant determined that there were no acute or subacute effects noted in the patient, but recommended that a pediatric gastroenterologist monitor the patient for cancer for an extended period of time.

Cause(s) — The event was caused by human error. The hospital staff member did not look at the dosage label before administering the radiopharmaceutical.

Actions Taken to Prevent Recurrence — Corrective actions taken by the licensee involved revision of their procedures to require dual verification of all dosages to be administered to children and retraining the staff on the new procedures.

This event is closed for the purpose of this report.

****

05-03 Medical Event at St. Joseph Regional Medical Center in South Bend, Indiana

Criterion IV, "For Medical Licensees," of Appendix A to this report states, in part, that a medical event that results in a dose that is (1) equal to or greater than 1 Gy (100 rads) to a major portion of the bone marrow, to the lens of the eye, or to the gonads or (2) equal to or greater than 10 Gy (1,000 rads) to any other organ; and represents a prescribed dose or dosage that is delivered to the wrong treatment site will be considered for reporting as an AO.

Date and Place —Between January 26 and March 22, 2004 (reported March 25, 2005 due to a misinterpretation of reporting requirements by the licensee), South Bend, Indiana

Nature and Probable Consequences — The licensee reported in March and April 2005, that between January 26 and March 22, 2004, three patients received unintended radiation doses to the skin of their thighs from cesium-137 brachytherapy sources. The vaginal applicator used for the treatments was loaded with incorrectly sized cesium-137 sources, which migrated from the intended treatment position through the placement spring when the patient moved to a more up-right position. As a result of the sources moving, the patient's inner thighs received unintended doses of radiation. Approximately two weeks after treatment, the patients developed skin lesions on their inner thighs. The licensee determined that these patients received unintended doses to a small area of the skin on the upper thigh of approximately 2000, 1500, and 2000 cGy (rad), respectively. Based on clinical observations, the licensee determined that all patients received the respective prescribed doses to the intended treatment areas. The referring physician and patients were notified of the event. The licensee referred the patients to other institutions and care providers for specialized followup wound care to treat the recurring skin ulcerations. The NRC retained a medical consultant during the inspection associated with the event. The long-term health effects on the patients, as a result of the unintended doses, is unknown.

Cause(s) — The causes of these events were improper source selection, inadequate manufacturer instructions, inadequate management oversight, and inadequate procedures.

Actions Taken to Prevent Recurrence — Corrective actions taken by the licensee involved modifying the applicator by using different hardware to hold the sources in place, revising their procedures, and retraining the staff on the new procedures.

This event is closed for the purpose of this report.

AGREEMENT STATE LICENSEES

During this reporting period, six events at Agreement State-licensed facilities were significant enough to be reported as AOs based on the criteria in Appendix A to this report.

AS 05-01   Iridium-192 Brachytherapy Seed Medical Event at LDS Hospital in Salt Lake City, Utah

Criterion IV, "For Medical Licensees," of Appendix A to this report states, in part, that a medical event that results in a dose that is (1) equal to or greater than 1 Gy (100 rads) to a major portion of the bone marrow, to the lens of the eye, or the gonads, or (2) equal to or greater than 10 Gy (1,000 rads) to any other organ; and represents a prescribed dose or dosage that is delivered to the wrong treatment site, will be considered for reporting as an AO.

Date and Place — October 26, 2004; LDS Hospital; Salt Lake City, Utah

Nature and Probable Consequences — A patient received 27.56 Gy (2,756 rads) instead of the prescribed 5 Gy (500 rads) during a high dose-rate (HDR) treatment for larynx cancer. The event involved an iridium-192 (Ir-192) source with an activity of 244.2 GBq (6.6 Ci). The error was caused by the use of the diameter instead of the radius of a circular tool to mark the treatment site in a computer software program. As a result, the area treated was 2 centimeters (cm) away from the intended treatment site. The error was discovered before the third fraction. The prescribing physician stopped the treatment until dosimetry information was completed. The licensee notified the patient and the patient's referring physician of the event. The licensee determined that the impact of the additional dose is probable acute radiation effects and possible late or chronic toxicities.

Cause(s) — This event was caused by human error. The incorrect size button corresponding to the circle tool was used, which caused the diameter instead of the radius to be used in the dosing plan. This caused the incorrect dose to be administered to the incorrect location.

Actions Taken to Prevent Recurrence

Licensee — The licensee suggested that the software manufacturer print the word "RADIUS" on the "size" button located adjacent to the circle tool. To date, the manufacturer has not responded to this issue. The licensee will measure the distance on the brachytherapy device's hard copy output with a ruler to confirm that the distance is entered correctly. The licensee also modified the HDR dose check program so that, in addition to confirming the doses to coordinates entered into the device's input, user specified point coordinates may be manually entered into the check program and compared to what is calculated.

State Agency — The Utah Division of Radiation Control investigated the event on November 3, 2004 and approved the corrective actions that the licensee implemented to prevent the recurrence.

This event is closed for the purpose of this report.

****

AS 05-02   Diagnostic Medical Event at Baystate Health Systems in Springfield, Massachusetts

Criterion IV, "For Medical Licensees," of Appendix A to this report states, in part, that a medical event that results in a dose that is (1) equal to or greater than 1 Gy (100 rads) to a major portion of the bone marrow, to the lens of the eye, or the gonads, or (2) equal to or greater than 10 Gy (1,000 rads) to any other organ; and represents a prescribed dose or dosage that is delivered by the wrong treatment mode, will be considered for reporting as an AO.

Date and Place — January 7, 2005; Baystate Health Systems; Springfield, Massachusetts

Nature and Probable Consequences — The licensee reported that a patient should have received 0.63 MBq (0.017 mCi) of iodine-131 (I-131) for a thyroid uptake study but instead received 133.2 MBq (3.6 mCi) of I-131 for a total body scan. A nuclear medicine technologist incorrectly placed the order for a total body scan instead of a thyroid uptake study without looking at the diagnosis. The I-131 was administered and it was later discovered that the wrong procedure was administered. The administration resulted in a thyroid dose of 131 Gy (13,100 rads). The patient and referring physician were notified of the error. The licensee indicated there would be no negative health effects from this administration because the patient had hyperthyroidism, thus, the unintended thyroid dose will be taken into account when additional I-131 is given to the patient.

Cause(s) — Human error in that the procedure was erroneously posted as a total body scan when it was actually a thyroid uptake study. This caused the wrong quantity of I-131 to be administered.

Actions Taken to Prevent Recurrence

Licensee — Corrective actions taken by the licensee involved modifying procedures to include removing Central Booking from radioisotope ordering (the referring physician will fax the order directly to Nuclear Medicine), switching from I-131 to I-123 for thyroid uptake studies, and revising the nuclear medicine request form for thyroid procedures.

State Agency — The State reviewed and approved the corrective actions taken by the licensee and will follow-up at the next inspection.

This event is closed for the purpose of this report.

5

AS 05-03   High Dose-Rate Afterloader Medical Event at Saddleback Memorial Medical Center in Laguna Hills, California

Criterion IV, "For Medical Licensees," of Appendix A to this report states, in part, that a medical event that results in a dose that is (1) equal to or greater than 1 Gy (100 rads) to a major portion of the bone marrow, to the lens of the eye, or the gonads, or (2) equal to or greater than 10 Gy (1,000 rads) to any other organ; and represents a prescribed dose or dosage that is delivered to the wrong treatment site will be considered for reporting as an AO.

Date and Place — January 24-28, 2005; Saddleback Memorial Medical Center; Laguna Hills, California

Nature and Probable Consequences — A patient undergoing therapeutic radiation treatment following a breast lumpectomy was treated with a high dose-rate (HDR) device using an iridium-192 (Ir-192) source with an activity of 277.5 GBq (7.5 Ci). The prescribed dose was 35 Gy (3,500 rads) to the inside of the breast at the site of the excised tumor, but instead the patient received 70 Gy (7,000 rads) to other portions of the breast during treatment. The unintended irradiation occurred when the HDR device was mispositioned. Re-evaluation of the treatment plan revealed that the wrong source wire travel distance was used during the treatment. The Ir-192 source was positioned 8 centimeters (cm) short of the planned location. The licensee believes the error occurred when the source wire travel distance was input to the HDR device; however, since no record was maintained of the source wire travel distance measured by the therapy technologist, this could not be verified. It is known that the incorrect distance was input to the HDR planning system. The patient and the referring physician were notified of the event. No long-term health effects are expected due to the unplanned tissue dose.

Cause(s) — This event was attributed to human error and an inadequate procedure.

Actions Taken to Prevent Recurrence

Licensee — A procedure was developed specifying the need to verify and document the verification of source wire travel distance determination and training on the correct input to the treatment planning system was performed. In addition, nominal source wire travel distances for expected types of HDR usage were added to the form utilized for recording the HDR treatment quality assurance checklist, thus providing a check on the determination of this parameter.

State Agency — State inspectors investigated the medical event and issued written violations for failure to follow a license condition that required independent verification of HDR treatment data input, and for failure to report the medical event to the state within 24 hours of its discovery. The State reviewed the licensee's corrective actions and found them adequate to prevent recurrence.

This event is closed for the purpose of this report.

AS 05-04   Yttrium-90 Therapeutic Medical Event at University of Wisconsin in Madison, Wisconsin

Criterion IV, "For Medical Licensees," of Appendix A to this report states, in part, that a medical event that results in a dose that is (1) equal to or greater than 1 Gy (100 rads) to a major portion of the bone marrow, to the lens of the eye, or the gonads, or (2) equal to or greater than 10 Gy (1,000 rads) to any other organ; and represents a prescribed dose or dosage that is delivered to the wrong treatment site will be considered for reporting as an AO.

Date and Place — April 5, 2005; University of Wisconsin in Madison; Madison, Wisconsin

Nature and Probable Consequences — A patient was administered a 1.78 GBq (48 mCi) dose of yttrium-90 (Y-90), instead of the intended 1.04 GBq (28 mCi) Y-90 dose. As a result of the medical event, the patient received a dose of 1.07 to 3.20 Gy (107 to 320 rads) to the red bone marrow, with a median exposure of 2.31 Gy (231 rads) from Y-90. The error was discovered on April 7, 2005, during a licensee review of records. The patient and referring physician were notified of the event. The licensee indicated there will be no negative health effects from this administration.

Cause(s) — Lack of management oversight which attributed to failure to prepare a written directive prior to the administration, a poor training program, and human error.

Actions Taken to Prevent Recurrence

Licensee — The licensee suspended the use of Y-90 and conducted a root cause investigation of the event. The licensee's corrective actions included writing new policies and procedures, implementing new training programs, and hiring new personnel.

State Agency — The State of Wisconsin investigated the event on April 11, 2005 and determined that the licensee (1) failed to prepare a written directive prior to administering the Y-90, (2) failed to prevent usage of a dose that differed from the intended dosage by more than 20 percent, (3) failed to establish appropriate administrative procedures, (4) failed to ensure radiation safety activities were performed under approved procedures, and (5) failed to instruct individuals working under the supervision of an authorized user of the licensee's written directive procedures. A medical consultant contracted by the State of Wisconsin determined that no adverse medical effects occurred as a result of this medical event. As a result of the State's investigation, the licensee implemented the corrective actions detailed above. The State reviewed the licensee's corrective actions and found them adequate to prevent recurrence.

This event is closed for the purpose of this report.

AS 05-05   Therapeutic Medical Event at University of Utah in Salt Lake City, Utah

Criterion IV, "For Medical Licensees," of Appendix A to this report states, in part, that a medical event that results in a dose that is (1) equal to or greater than 1 Gy (100 rads) to a major portion of the bone marrow, to the lens of the eye, or the gonads, or (2) equal to or greater than 10 Gy (1,000 rads) to any other organ; and represents a prescribed dose or dosage that is delivered to the wrong treatment site, will be considered for reporting as an AO.

Date and Place — August 4, 2005; University of Utah; Salt Lake City, Utah

Nature and Probable Consequences — A patient received radiation therapy to the left bronchus using a high dose-rate (HDR) device. The HDR contained a 252 GBq (6.81 Ci) iridium-192 (Ir-192) source. The prescribed radiation therapy treatment plan called for three treatments to the left bronchus, each fraction to deliver a dose of 7 Gy (700 rads). The medical event, which occurred during the second treatment, was due to a 3-centimeter (cm) error in the source wire travel distance. The source wire distance was entered incorrectly by a medical physicist. As a result, a 3 cm length of the left bronchus received approximately 6.40 to 18.60 Gy (640 to 1,860 rads) at a 0.5 cm depth and 2.54 to 6.62 Gy (254 to 662 rads) at a 1 cm depth. A 3-cm region next to the intended treatment site received up to 6 Gy (600 rads) less than the prescribed dose. The licensee notified the patient and the patient's referring physician of the event. The patient received no adverse health effects from the medical event.

Cause(s) — This event was attributed to human error in that the treatment site was not verified.

Actions Taken to Prevent Recurrence

Licensee — The licensee implemented a new procedure adding a question to verify the treatment distances during HDR treatments.

State Agency — The State has reviewed and accepted the licensee's corrective actions.

This event is closed for the purpose of this report.

AS 05-06   Dose to Fetus at Riverside Methodist Hospital in Columbus, Ohio

Criterion I.A.2, "For All Licensees," of Appendix A to this report states, "Any unintended radiation exposure to any minor (an individual less than 18 years of age) resulting in an annual total effective dose equivalent (TEDE) of 50 millisieverts (mSv) (5 rem) or more, or to an embryo/fetus resulting in a dose equivalent of 50 mSv (5 rem) or more," will be considered for reporting as an AO.

Date and Place — November 2 and November 16, 2004; Riverside Methodist Hospital; Columbus, Ohio

Nature and Probable Consequences — On November 2, 2004, a patient was administered 7.59 MBq (0.205 mCi) of iodine-123 (I-123) as part of a diagnostic procedure for hyperthyroidism. On November 16, 2004, the patient returned for a therapeutic treatment and was administered 469.9 MBq (12.7 mCi) of iodine-131 (I-131) as treatment. Prior to this administration, the patient was counseled regarding pregnancy and acknowledged, in writing, that she was not and could not be pregnant at that time. A pregnancy test was not performed to confirm this declaration. Later, the patient saw her physician because of abdominal pain. A radiograph of the abdomen revealed the pregnancy. A prenatal specialist determined that the fetus was 17 weeks old at the time of the I-131 administration. The dose estimate for the fetus was 0.024 Gy (2.04 rads) to the whole body and 224 Gy (22,400 rads) to the fetal thyroid from both I-123 and I-131 administrations. The perinatal specialist performed a blood test on the fetus and confirmed that the fetus had hyperthyroidism. An ultrasound test on the fetus showed no abnormalities in fetal development. The perinatal specialist will perform treatments in-utero to mitigate the effects of hyperthyroidism. The referring physician and patient were notified of the medical event.

Cause(s) - The cause of the event was human error. At the time of the administration, the patient was unaware of her pregnancy status and completed forms indicating that she was not pregnant.

Actions Taken to Prevent Recurrence

Licensee — The licensee has implemented a policy performing a serum pregnancy test and receiving the results within 80 hours of administration of therapeutic amounts of I-131. This test will be performed on all women 13 to 50 years of age, unless the women have been surgically sterilized.

State Agency — The Ohio Department of Health performed an on-site investigation on January 28, 2005 and determined that the licensee followed all required procedures. The State agency will conduct periodic inspections to ensure that the licensee's actions taken to prevent recurrence were implemented.

This event is closed for the purpose of this report.

ABNORMAL OCCURRENCE CRITERIA AND GUIDELINES FOR OTHER EVENTS OF INTEREST

An accident or event will be considered an AO if it involves a major reduction in the degree of protection of public health or safety. This type of incident or event would have a moderate or more severe impact on public health or safety and could include, but need not be limited to, the following:

(1)   Moderate exposure to, or release of, radioactive material licensed by or otherwise regulated by the Commission;

(2)   Major degradation of essential safety-related equipment; or

(3)   Major deficiencies in design, construction, use of, or management controls for facilities or radioactive material licensed by or otherwise regulated by the Commission.

The following criteria for determining an AO and the guidelines for "Other Events of Interest" were stated in an NRC policy statement published in the *Federal Register* on December 19, 1996 (61 FR 67072). The policy statement was revised to include criteria for gaseous diffusion plants and was published in the *Federal Register* on April 17, 1997 (62 FR 18820).

Note that in addition to the criteria for fuel cycle facilities (Section III of the AO criteria) that are applicable to licensees and certificate holders, such as the gaseous diffusion plants, other criteria that reference "licensees," "licensed facility," or "licensed material" also may be applied to events at facilities of certificate holders.

The guidelines for including events in Appendix C "Other Events of Interest" of this report were provided by the Commission in the Staff Requirements Memorandum on SECY-98-175, dated September 4, 1998, and are listed at the end of this appendix.

Abnormal Occurrence Criteria

Criteria by types of events used to determine which events will be considered for reporting as AOs are as follows:

I.   For All Licensees.

A.   Human Exposure to Radiation from Licensed Material

1.   Any unintended radiation exposure[2] to an adult (any individual 18 years of

---

[1]   An unintended radiation exposure for the purpose of reporting as an AO includes any occupational exposure, exposure to the general public, or exposure as a result of a medical event involving the wrong patient that exceeds the reporting values established in the regulation. All other reporting medical events will be considered for reporting as an AO under the criteria "For Medical Licensees".

In addition, unintended radiation exposures includes any exposure to a nursing infant, fetus, or embryo as a result of an exposure (other than an occupational exposure to an undeclared pregnant woman) to a nursing mother or pregnant woman.

age or older) resulting in an annual total effective dose equivalent (TEDE) of 250 mSv (25 rem) or more; or an annual sum of the deep dose equivalent (external dose) and committed dose equivalent (intake of radioactive material) to any individual organ other than the lens of the eye, bone marrow, and the gonads, of 2,500 mSv (250 rem) or more; or an annual dose equivalent to the lens of the eye, of 1 Sv (100 rem) or more; or an annual sum of the deep dose equivalent and committed dose equivalent to the bone marrow, and the gonads, of 1 Sv (100 rem) or more; or an annual shallow-dose equivalent to the skin or extremities of 2,500 mSv (250 rem) or more.

2.   Any unintended radiation exposure to any minor (an individual less than 18 years of age) resulting in an annual TEDE of 50 mSv (5 rem) or more, or to an embryo/fetus resulting in a dose equivalent of 50 mSv (5 rem) or more.

3.   Any radiation exposure that has resulted in unintended permanent functional damage to an organ or a physiological system as determined by a physician.

B.   Discharge or Dispersal of Radioactive Material from its Intended Place of Confinement

1.   The release of radioactive material to an unrestricted area in concentrations which, if averaged over a period of 24 hours, exceeds 5,000 times the values specified in Table 2 of Appendix B to 10 CFR Part 20, unless the licensee has demonstrated compliance with § 20.1301 using § 20.1302 (b) (1) or § 20.1302 (b) (2) (ii).

2.   Radiation levels in excess of the design values for a package, or the loss of confinement of radioactive material resulting in one or more of the following: (a) a radiation dose rate of 10 mSv (1 rem) per hour or more at 1 meter (3.28 feet) from the accessible external surface of a package containing radioactive material; (b) a radiation dose rate of 50 mSv (5 rem) per hour or more on the accessible external surface of a package containing radioactive material and that meet the requirements for "exclusive use" as defined in 10 CFR 71.47; or (c) release of radioactive material from a package in amounts greater than the regulatory limits in 10 CFR 71.51(a)(2).

C.   Theft, Diversion, or Loss of Licensed Material, or Sabotage or Security Breach[3]

---

2   Information pertaining to certain incidents may be either classified or under consideration for classification because of national security implications. Classified information will be withheld when formally reporting these incidents in accordance with Section 208 of the ERA of 1974, as amended. Any classified details regarding these incidents would be available to the Congress, upon request, under appropriate security arrangements.

1. Any lost, stolen, or abandoned sources that exceed 0.01 times the A_1 values, as listed in 10 CFR Part 71, Appendix A, Table A-1, for special form (sealed/nondispersible) sources, or the smaller of the A_2 or 0.01 times the A_1 values, as listed in Table A-1, for normal form (unsealed/dispersible) sources or for sources for which the form is not known. Excluded from reporting under this criterion are those events involving sources that are lost, stolen, or abandoned under the following conditions: sources abandoned in accordance with the requirements of 10 CFR 39.77(a); sealed sources contained in labeled, rugged source housings; recovered sources with sufficient indication that doses in excess of the reporting thresholds specified in AO criteria I.A.1 and I.A.2 did not occur during the time the source was missing; and unrecoverable sources lost under such conditions that doses in excess of the reporting thresholds specified in AO criteria I.A.1 and I.A.2 were not known to have occurred.

2. A substantiated case of actual or attempted theft or diversion of licensed material or sabotage of a facility.

3. Any substantiated loss of special nuclear material or any substantiated inventory discrepancy that is judged to be significant relative to normally expected performance, and that is judged to be caused by theft or diversion or by substantial breakdown of the accountability system.

4. Any substantial breakdown of physical security or material control (i.e., access control containment or accountability systems) that significantly weakened the protection against theft, diversion, or sabotage.

D. Other Events (i.e., Those Concerning Design, Analysis, Construction, Testing, Operation, Use, or Disposal of Licensed Facilities or Regulated Materials)

1. An accidental criticality [10 CFR 70.52(a)].

2. A major deficiency in design, construction, control, or operation having significant safety implications requiring immediate remedial action.

3. A serious deficiency in management or procedural controls in major areas.

4. Series of events (where individual events are not of major importance), recurring incidents, and incidents with implications for similar facilities (generic incidents) that create a major safety concern.

II.   For Commercial Nuclear Power Plant Licensees

    A.   Malfunction of Facility, Structures, or Equipment

        1.   Exceeding a safety limit of license technical specification (TS) [10 CFR 50.36(c)].

        2.   Serious degradation of fuel integrity, primary coolant pressure boundary, or primary containment boundary.

        3.   Loss of plant capability to perform essential safety functions so that a release of radioactive materials, which could result in exceeding the dose limits of 10 CFR Part 100 or 5 times the dose limits of 10 CFR Part 50, Appendix A, General Design Criterion (GDC) 19, could occur from a postulated transient or accident (e.g., loss of emergency core cooling system, loss of control rod system).

    B.   Design or Safety Analysis Deficiency, Personnel Error, or Procedural or Administrative Inadequacy

        1.   Discovery of a major condition not specifically considered in the safety analysis report (SAR) or TS that requires immediate remedial action.

        2.   Personnel error or procedural deficiencies that result in loss of plant capability to perform essential safety functions so that a release of radioactive materials, which could result in exceeding the dose limits of 10 CFR Part 100 or 5 times the dose limits of 10 CFR Part 50, Appendix A, GDC 19, could occur from a postulated transient or accident (e.g., loss of emergency core cooling system, loss of control rod system).

III.   For Fuel Cycle Facilities

    A.   A shutdown of the plant or portion of the plant resulting from a significant event and/or violation of a law, regulation, or a license/certificate condition.

    B.   A major condition or significant event not considered in the license/certificate that requires immediate remedial action.

    C.   A major condition or significant event that seriously compromises the ability of a safety system to perform its designated function that requires immediate remedial action to prevent a criticality, radiological, or chemical process hazard.

IV.   For Medical Licensees

A medical event that:

A.   Results in a dose that is (1) equal to or greater than 1 Gy (100 rads) to a major portion of the bone marrow, to the lens of the eye, or to the gonads, or (2) equal to or greater than 10 Gy (1,000 rads) to any other organ; and

B.   Represents either (1) a dose or dosage that is at least 50 percent greater than that prescribed in a written directive or (2) a prescribed dose or dosage that (i) is the wrong radiopharmaceutical,[4] or (ii) is delivered by the wrong route of administration, or (iii) is delivered to the wrong treatment site, or (iv) is delivered by the wrong treatment mode, or (v) is from a leaking source or sources.

Guidelines for "Other Events of Interest"

The Commission may determine that events other than AOs may be of interest to Congress and the public and should be included in an appendix to the AO report as "Other Events of Interest." Guidelines for events to be included in the AO report for this purpose may include, but not necessarily be limited to, events that do not meet the AO criteria but that have been perceived by Congress or the public to be of high health and safety significance, have received significant media coverage, or have caused the NRC to increase its attention to or oversight of a program area, or a group of similar events that have resulted in licensed materials entering the public domain in an uncontrolled manner.

---

3   "The wrong radiopharmaceutical" as used in the AO criterion for a medical event refers to any radiopharmaceutical other than the one listed in the written directive or in the clinical procedures manual.

APPENDIX B
UPDATES OF PREVIOUSLY REPORTED ABNORMAL OCCURRENCES

During this reporting period, no new significant information became available regarding any AO event that the NRC previously reported in the FY 2004 Report to Congress on Abnormal Occurrences.

APPENDIX C
OTHER EVENTS OF INTEREST

This appendix discusses "Other Events of Interest" that do not meet the abnormal occurrence (AO) criteria in Appendix A, but have been perceived by Congress or the public to be of high health and safety significance, have received significant media coverage, or have caused the NRC to increase its attention to or oversight of a program area, including a group of similar events that have resulted in licensed materials entering the public domain in an uncontrolled manner.

NUCLEAR POWER PLANTS

1.   <u>Safe Shutdown Potentially Challenged Due To Unanalyzed Internal Flooding Events and Inadequate Design at Kewaunee Power Station</u>

The following event did not meet the AO criteria because it did not involve a major reduction in the protection of public health or safety. However, the event involved an issue with substantial importance to safety. This event is being included in this year's report because it received significant media coverage.

In March 2005, with the Kewaunee Power Station (KPS) shut down and in an extended outage, the licensee reported an NRC-identified finding concerning the vulnerability to a design-basis internal flood of multiple trains of safety-related systems necessary for safe-shutdown.

At KPS, multiple trains of safety-related systems necessary for safe-shutdown (e.g., auxiliary feedwater, emergency diesel generators, and electrical distribution switchgear) are located at the same elevation and immediately adjacent to the turbine building basement. During inspections of the turbine building and adjacent spaces, NRC inspectors identified that water from a turbine building flood could have flowed into these spaces through non-watertight doors and through the floor drain system, which consisted of an open pipe connecting the spaces to the turbine building sump. Several non-safety related systems are located within the turbine building that have the potential to release large volumes of water in the event of a pressure boundary failure. These include the circulating water, condensate storage, and fire protection systems. In the case of the fire protection system, a large volume of water could be released simply from system actuation. Although infrequent, U.S. nuclear power plant operating experience includes internal flooding events initiated by circulating water expansion joint failure and fire protection system pressure boundary failure. Had a significant internal flooding event occurred in the KPS turbine building basement, the safe shut down would have been substantially challenged.

The root cause for this issue was the failure of the licensee to conduct adequate engineering analysis of the facility to ensure the plant's design basis was met. Contributing causes included a lack of complete understanding of the design basis by the licensee; a lack of a complete understanding of the risk associated with internal flooding events by the licensee; the failure to adequately evaluate and implement actions to address industry operating experience; and the failure to adequately resolve known deficiencies.

In February 2005, the licensee shut down KPS and entered an extended outage to address these and other design concerns. Corrective actions were taken during the approximately four month outage to ensure safety-related equipment will be adequately protected against postulated failures

16

of non-safety related piping systems, including high energy line breaks, random pipe failures, and seismically induced pipe failures. These corrective actions include the compilation of design and licensing bases for internal flooding to support current and future flooding design; seismic qualification of selected piping and components; design modifications to protect safety related systems necessary for safe-shutdown including the installation of check valves in selected floor drains; auxiliary feedwater pump lube oil cooler and drain flowpath revisions; and installation of a circulating water pump trip on high TB basement water level, flood barriers at doors to safety-related equipment rooms, and enhanced supports for auxiliary feedwater pump steam supply piping.

The NRC inspected the above modifications that addressed the plant deficiencies.

The NRC staff evaluated and provisionally rated the KPS event as Level 2 (Incident) on the International Nuclear Event Scale and posted it on the Nuclear Events Web-based System. On November 7, 2005, the NRC staff issued an Information Notice (IN 2005-30) to all holders of operating licenses for nuclear power reactors to alert licensees to the importance of maintaining the plant flooding analysis and design and ensure that internal flooding risk is effectively managed consistent with NRC requirements and principles of effective risk management.

On December 21, 2005, the NRC issued a NOV for a violation involving the failure to provide adequate design control to ensure that the design of KPS prevented turbine building flooding from impacting multiple safety related equipment trains needed to achieve and maintain safe shutdown of the plant. This inspection finding was characterized as Yellow (i.e., an issue with substantial importance to safety, that will result in additional NRC inspection and potentially other NRC action).

This event is closed for the purpose of this report.

****

2.   Missing Fuel Rod Segments at Humboldt Bay Power Plant

This event did not meet the AO criteria because it did not involve a major reduction in the degree of protection of public health and safety. Nonetheless, this event was included in the FY 2004 *report to Congress on Abnormal Occurrences* because it received significant public interest. An update of this event is provided.

On July 16, 2004, Pacific Gas and Electric Company (PG&E) (the licensee), notified the NRC of a discrepancy between inventory records and the physical location of three spent fuel rod segments, each approximately 18-inches long, that were previously known to be at the Humboldt Bay Power Plant. The licensee submitted a 30-day follow-up report pursuant to 10 CFR 20.2201 (b)(2)(ii) on August 16, 2004. The licensee searched for the segments in the most likely and issued on August 19, 2005. The team determined that PG&E's (the licensee's) investigation was thorough and complete, and the conclusions were reasonable and supportable. The NRC also concluded that once the licensee made discoveries about the missing rod segments, the NRC was promptly and accurately notified and kept informed throughout the process of search and investigation.

The NRC staff concluded that the current material control and accounting (MC&A) program being implemented by the licensee meets regulatory requirements. The NRC team found no evidence to support theft or diversion of the missing spent fuel. The NRC team concluded that the most likely scenario for the missing spent fuel is that is was inadvertently shipped to a low level waste burial site. While the licensee's conclusions supported this as a possibility, the licensee concluded that the most likely scenario for the missing spent fuel rod segments was that they are still onsite in the spent fuel pool in an altered condition.

The NRC's final inspection report identified three apparent violations. The apparent violations involved (1) the failure to keep records showing the inventory, transfer or disposal of the three 18-inch segments of irradiated fuel and one complete and three partial core detectors; (2) the failure to establish, maintain, and follow adequate written material control and accounting procedures sufficient to account for the special nuclear material (SNM) contained in the fuel rod segments; and (3) the failure to conduct an accurate physical inventory of all SNM in the licensee's possession at intervals that do not exceed 12 months.

These three violations were cited in a Notice of Violation issued on December 20, 2005. Because the loss of control of highly radioactive SNM is a very significant concern to the NRC, the three violations were categorized collectively as a Severity Level II problem in accordance with the NRC's Enforcement Policy. To emphasize the importance of controlling SNM and of prompt identification of violations, the NRC issued a Proposed Imposition of Civil Penalty in the amount of $96,000 for the Severity Level II problem.

The NRC has issued generic communications to other licensees informing them of previously identified problems and requesting them to examine their MC&A programs and account for their fuel. The NRC is conducting inspections of MC&A programs at all nuclear power plants and evaluating the information gathered during the inspections.

This event is closed for the purpose of this report.

NRC FORM 335 (9-2004) NRCMD 3.7	U.S. NUCLEAR REGULATORY COMMISSION	1. REPORT NUMBER (Assigned by NRC, Add Vol., Supp., Rev., and Addendum Numbers, if any.)
	BIBLIOGRAPHIC DATA SHEET *(See instructions on the reverse)*	NUREG-0090, Vol. 28

2. TITLE AND SUBTITLE	3. DATE REPORT PUBLISHED	
Report to Congress on Abnormal Occurrences, Fiscal Year 2005	MONTH	YEAR
	April	2006
	4. FIN OR GRANT NUMBER	

5. AUTHOR(S)	6. TYPE OF REPORT
	Annual
	7. PERIOD COVERED *(Inclusive Dates)* Fiscal Year 2005

8. PERFORMING ORGANIZATION - NAME AND ADDRESS *(If NRC, provide Division, Office or Region, U.S. Nuclear Regulatory Commission, and mailing address; if contractor, provide name and mailing address.)*

Division of Fuel, Engineering and Radiological Research (DFERR)
Office of Nuclear Regulatory Research
U. S. Nuclear Regulatory Commission
Washington, DC 20555-0001

9. SPONSORING ORGANIZATION - NAME AND ADDRESS *(If NRC, type "Same as above"; if contractor, provide NRC Division, Office or Region, U.S. Nuclear Regulatory Commission, and mailing address.)*

Same as 8., above

10. SUPPLEMENTARY NOTES

11. ABSTRACT *(200 words or less)*

Section 208 of the Energy Reorganization Act of 1974 identifies an abnormal occurrence (AO) as an unscheduled incident or event that the Nuclear Regulatory Commission (NRC) determines to be significant from the standpoint of public health or safety. The Federal Reports Elimination and Sunset Act of 1995 requires that AOs be reported to Congress on an annual basis. This report includes those events that the NRC has determined to be AOs during fiscal year 2005.

This report addresses nine AOs. Three of these events occurred at facilities licensed by the NRC and six events involved Agreement State licensees.

12. KEY WORDS/DESCRIPTORS *(List words or phrases that will assist researchers in locating the report.)*	13. AVAILABILITY STATEMENT
Exposure, Dose, Brachytherapy, Medical Event	unlimited
	14. SECURITY CLASSIFICATION
	(This Page) unclassified
	(This Report) unclassified
	15. NUMBER OF PAGES
	16. PRICE

NRC FORM 335 (9-2004)   PRINTED ON RECYCLED PAPER

www.ingramcontent.com/pod-product-compliance
Lightning Source LLC
Chambersburg PA
CBHW081416170526
45166CB00010B/3364